I0471964

Manufacturing Mill Fire
Methuen, Massachusetts

Investigated by: Scott M. Howell
Edited by: J. Gordon Routley

This is Report 110 of the Major Fires Investigation Project conducted by Varley-Campbell and Associates, Inc. under contract EMW-94-C-4423 to the United States Fire Administration, Federal Emergency Management Agency.

Department of Homeland Security
United States Fire Administration
National Fire Data Center

U.S. Fire Administration Fire Investigations Program

The U.S. Fire Administration develops reports on selected major fires throughout the country. The fires usually involve multiple deaths or a large loss of property. But the primary criterion for deciding to do a report is whether it will result in significant "lessons learned." In some cases these lessons bring to light new knowledge about fire--the effect of building construction or contents, human behavior in fire, etc. In other cases, the lessons are not new but are serious enough to highlight once again, with yet another fire tragedy report. In some cases, special reports are developed to discuss events, drills, or new technologies which are of interest to the fire service.

The reports are sent to fire magazines and are distributed at National and Regional fire meetings. The International Association of Fire Chiefs assists the USFA in disseminating the findings throughout the fire service. On a continuing basis the reports are available on request from the USFA; announcements of their availability are published widely in fire journals and newsletters.

This body of work provides detailed information on the nature of the fire problem for policymakers who must decide on allocations of resources between fire and other pressing problems, and within the fire service to improve codes and code enforcement, training, public fire education, building technology, and other related areas.

The Fire Administration, which has no regulatory authority, sends an experienced fire investigator into a community after a major incident only after having conferred with the local fire authorities to insure that the assistance and presence of the USFA would be supportive and would in no way interfere with any review of the incident they are themselves conducting. The intent is not to arrive during the event or even immediately after, but rather after the dust settles, so that a complete and objective review of all the important aspects of the incident can be made. Local authorities review the USFA's report while it is in draft. The USFA investigator or team is available to local authorities should they wish to request technical assistance for their own investigation.

This report and its recommendations were developed by USFA staff and by Varley-Campbell & Associates, Inc., Miami and Chicago, its staff and consultants, who are under contract to assist the Fire Administration in carrying out the Fire Reports Program.

The U.S. Fire Administration greatly appreciates the cooperation and information received from the Methuen Fire Department, most particularly Fire Chief Kenneth Bourassa and Inspector Glenn Gallant, and Detective Lieutenant Robert Corry of the Massachusetts State Police.

For additional copies of this report write to the U.S. Fire Administration, 16825 South Seton Avenue, Emmitsburg, Maryland 21727. The report is available on the Administration's Web site at http://www.usfa.dhs.gov/

U.S. Fire Administration

Mission Statement

As an entity of the Department of Homeland Security, the mission of the USFA is to reduce life and economic losses due to fire and related emergencies, through leadership, advocacy, coordination, and support. We serve the Nation independently, in coordination with other Federal agencies, and in partnership with fire protection and emergency service communities. With a commitment to excellence, we provide public education, training, technology, and data initiatives.

TABLE OF CONTENTS

Manufacturing Mill Fire
Methuen, Massachusetts
December 11, 1995

Investigated By: Scott M. Howell
Edited by: J. Gordon Routley

Local Contacts: Kenneth F. Bourassa, Fire Chief
Glenn A. Gallant, Inspector

Methuen Fire Department
24 Lowell Street
Methuen, MA 01844
508-794-3252

Det. Lt. Robert A. Corry, Fire Investigator
Massachusetts State Police

OVERVIEW

An explosion and fire in an industrial complex in Methuen, Massachusetts on December 11, 1995 injured 37 people and destroyed nearly one million square feet of manufacturing space. Several workers were critically burned by the initial explosion and six firefighters received minor injuries as the incident progressed. The ensuing fire, driven by 40 mile per hour winds, destroyed several large buildings in the complex, which straddled the municipal boundary between the cities of Lawrence and Methuen.

Malden Mills, a manufacturer of insulated fabric for winter clothing, was the largest employer in the area and the fire put about 1700 people out of work. The damage estimate stands at $500 million, the largest property damage fire loss in the history of the Commonwealth of Massachusetts. Fire brands ignited a fire in a furniture refinishing company more than one quarter of a mile away.

The fire required the evacuation of over 60 neighborhood homes for several days. Floating embers ignited several spot fires.

In all, about 55 area fire and EMS agencies responded to the incident with more than 400 personnel and 100 pieces of apparatus. The fire took over twelve hours to bring under control.

1

KEY ISSUES

Issues	Comments
Construction	The property was a large complex of interconnected mill construction buildings with narrow spaces between the major structures.
Weather Conditions	Operations were conducted in winter conditions with temperatures below 10 degrees Fahrenheit and 40 mph winds driving the flames. Frozen hydrants and high winds hampered the fire suppression efforts within the mill complex.
Fire Origin	An explosion of unknown origin ignited the fire in a processing area, inside a large manufacturing building.
Search & Rescue	The Methuen Fire Department performed search and rescue in more than 56,000 square feet of mill floor space with a very limited number of personnel on the scene. The need for search and rescue operations delayed fire discovery for about 15 minutes.
Fire Discovery	A waterflow alarm was transmitted without delay. However the location of the fire was not discovered until search and rescue operations had been completed.
Automatic Sprinklers	The sprinkler system was overtaxed early due to the number of heads activated by the initial explosion. Some of the sprinkler piping may have been damaged by the explosion.
Casualties	27 civilian injuries, including 12 with critical burns, and 6 firefighter injuries.
Exterior Storage	Semi-trailers, which were used for additional storage within the complex, blocked access to some areas and to yard hydrants. Some of the trailers became involved in the fire.
Mutual Aid	Mutual aid was requested immediately. More than 400 firefighters and units from more than 50 fire departments eventually responded.
Pre-Fire Planning	No pre-fire plan information was available. Mill employees had useful information and tried to help.
Defensive Strategy	Due to the size of the building, limited water supply and the rapid fire spread, defensive strategy was employed. The fire could not be stopped until it reached an open area where defensive forces could be assembled.

LOCATION

Methuen is a city with a population of 41,000, located about 25 miles north of Boston on the Massachusetts-New Hampshire border. It is one of several contiguous communities located in the Merrimac River Valley. The 24 square-mile city is similar to many of the older industrial cities in New England, with densely built manufacturing complexes directly adjacent to crowded residential areas. These communities have a high level of fire risk and have experienced many major fires in similar complexes over the past century.

The cities of Salem, New Hampshire and Lawrence, Massachusetts border Methuen on the north and south. The dividing line between Methuen and Lawrence bisects the Malden Mills complex and passes through the building where the fire originated, while the boundary between Massachusetts and New Hampshire is less than 2 miles north of the fire scene.

The Methuen Fire Department has 96 career personnel and operates four stations with a normal complement of 19 personnel on duty. An engine company with 3 crew members is assigned to each station, in addition to a rescue squad, a BLS ambulance and a ladder truck which all respond from the Central Fire Station. A Deputy Chief is in charge of each shift and has an assigned driver. In 1994 the department responded to 1,943 fire and 2,475 EMS calls.

Mutual aid is available from several adjacent communities, including Lawrence and Salem, as well as a large number of career and volunteer fire departments in the Merrimac Valley area. Automatic mutual aid responses are designated for multiple alarms throughout the region. The Haverhill Communications Center normally coordinates responses and coverage for incidents that go beyond a 4th alarm level.

Emergency medical service is operated as a two-tier system, with the Methuen Fire Department providing initial response and BLS ambulance service. Advanced life support is provided by hospital-based paramedic units, which are dispatched by CMED, a Regional communications and coordination center operated by Northeast Emergency Medical Services, a non-profit corporation. The ALS units respond to incidents when requested by the local jurisdiction and the paramedics accompany patients in the fire department ambulances. The ALS vehicles do not have transport capabilities.

THE COMPLEX

The original buildings in the complex occupied by Malden Mills Industries, Inc. were constructed in the late 1800's and early 1900's and are typical of heavy timber mill construction industrial buildings found in many New England towns and cities. These large multi-story buildings were constructed with brick and mortar exterior bearing walls and with heavy timber interior columns and floor joists. The interior floor areas were designed with large open floor areas and ceiling heights of up to 20 feet to allow for machinery and unobstructed flow of materials. Automatic sprinklers were provided to reduce the risk of a fire involving the interior of one of these large structures.

The complex included several large mill construction buildings, stretching for blocks along the banks of the Spickett River. It was originally constructed with separations of 50 to 60 feet between the major buildings, however numerous additions to accommodate the changing needs of the tenants had narrowed the alleys to less than 30 feet in some places. Several overhead passageways were added over the years to allow for the flow of materials from one area to another, interconnecting several of the buildings and providing potential paths for fire extension.[1]

The complex was originally constructed as a piano factory. From the 1920's to the 1940's, the property was occupied by a clothing manufacturer, until an exodus of the textile industry from the area left the complex vacant for several years. The property was subsequently subdivided among a number of owners as parts were converted and occupied by different tenants for a variety of purposes.

In 1956, Malden Mills Industries located a division on the site to take advantage of the location, facilities, and the glut of skilled workers left behind in the wake of the migration of the textile industry out of New England. The new occupants were very successful at developing new products and gradually expanded their operations into several of the buildings in the complex. At the time of the fire Malden Mills employed approximately 2,500 workers at this location. Other tenants, including Acadia Mills, occupied additional buildings within the complex.

Building Descriptions

The fire originated in the Monomac Building, which was approximately 500 feet long, 100 feet wide and 5 stories (approximately 100 feet) high. The Acadia Mills Building, which was about 360 x 100

---

[1] It was reported that fire doors were installed at each end of the passageways where they connected to the major buildings.

feet in area and three stories (60 feet) in height was located about 50 feet north of the Monomac Building. These two large buildings ran north and south, parallel to the Spickett River, and were connected by an enclosed overhead passageway at the third floor level. (Figure 1).

East of the Monomac Building was the Process Building, which was about 500 x 100 feet in area and four stories (80 feet) in height. An enclosed passage spanned from the middle of the Monomac Building to the south end of the Process Building connecting the 2nd, 3rd, and 4th floors. Another overhead passageway connected the 2nd floor of the Process Building to a two-story section of the Acadia Mills Building.

To the east of the Process Building, was the Main Mill Building. This structure was four stories (80 feet) high and covered an area of approximately 580 x 120 feet. The Dye House, a newer 30 x 580-foot single-story concrete block structure had been added along the east side of the Main Mill Building. The New Building, a 155 x 360-foot one-story concrete block structure, was connected to the north half of the Dye House.

The New Building bordered French and Chase Streets, which were the east and north perimeters of the Malden Mills property. Across these narrow streets the neighborhood consisted of two and three story wood frame residential buildings. Broadway Street, which is primarily commercial, is one block to the east. Stafford Street provided access from Broadway to the middle of the complex, south of the Main and Process Buildings. An elevated covered walkway, 385 feet long, crossed Stafford Street and a large employee parking area on the south side to connect the Process Building to the Marriner Building.

The Spickett River formed the west perimeter of the property. Between the buildings and the river was an open area, approximately 50 feet wide. A large number of 40 foot trailers, some of which were used for storage, were parked in this area. Several trailers were parked at the loading dock and more were parked along the west wall of the building.

The original mill complex extended north and south along the river bank and included several additional structures which were not involved in the fire. The dividing line between Methuen and Lawrence was located just south of Stafford Street and ran through the southern section of the Monomac Building.

Fire Protection Systems

Most of the buildings in the complex were protected by wet pipe automatic sprinklers[2] and standpipe systems and the larger buildings had multiple sprinkler systems covering different zones. The waterflow and manual alarms were connected to the Methuen fire alarm box system via an auxiliary circuit and were also monitored at an annunciator panel in the main guard station.

Water for the fire protection systems in the complex came from two separate sources. The sprinkler and standpipe systems in the Monomac, Process and Acadia Buildings and the yard hydrants surrounding these buildings were supplied by an on-site fire protection system. The sprinkler systems in the Main Building and the Dye House were connected to the City of Methuen municipal water system, while hose stations in the north and south stairwells were connected to the on-site system. There was no fire pump supporting the municipal water supply and there were no fire department

---

[2] There were automatic sprinklers in all of the major buildings that were involved in the fire, except the New Building.

connections that would have allowed for additional water supply to be provided to any of these systems from fire department pumpers.

From on-site inspections and familiarization tours, the fire department knew that the sprinkler system in the Monomac, Process and Acadia Mill Buildings and the yard hydrants were all interconnected, with no fire department connections on any buildings. If the sprinkler system in any one building was overpowered, then the sprinklers in the other two building would be crippled from the lack of water and the yard hydrant system would be similarly destroyed.

The on-site fire protection system was supplied by a 2000 gpm electric fire pump and a 2000 gpm diesel powered backup pump which both drew water from the river. Both pumps could operate simultaneously to supply a rated capacity of 4000 gallons per minute to the fire protection systems. A 750 gpm jockey pump was set to maintain a static pressure of 120 psi in this system. A 14 inch main delivered water from the pump house to the alley between the Monomac and Process Buildings. A 10 inch main was buried to the west of the Monomac and Acadia Buildings, parallel to the river, and the loop was completed by a 6 inch main. (Figure 3).

The 14 inch main in the alley between the Monomac and Process Buildings supplied water to the sprinklers in both buildings[3]. Each building had four 6 inch sprinkler risers which could be shut-off by underground valves located about 10 feet outside the exterior walls. Each riser supplied the sprinklers in an area of the first floor and the corresponding area on the upper floors. The individual zones were controlled by separate supervised OS&Y valves. The sprinkler lines in some unheated areas on the west side of the Monomac Building were filled with antifreeze sections. Hose stations were located in the stairways at each floor level.

The municipal water supply for the complex came from two connections to a 12 inch main under Broadway Street to the east of the complex. A 14 inch main entered the complex at Chase Street and a 6 inch line entered at Stafford Street. These lines supplied the water for the sprinklers in the Main Building and the Dye House. There was reported to be an interconnection between the municipal water system and the on-site fire protection system, however, the manually operated valve between the two systems was normally closed. This valve, located at the southeast corner of the Main Building, was not known to the fire department and was not opened during the fire.

There was also a closed connection between the Methuen and Lawrence municipal water systems. At one time each city provided water for the buildings within their respective municipalities. A more recent agreement between the property owner and the two cities called for Methuen to provide the public water supply for the entire complex. A street valve between the two systems was normally kept closed because they operated at different pressures; a static pressure of about 65 psi was normally maintained in the Lawrence system, while the Methuen system was maintained at about 110 psi.

During the fire investigation another underground line and valve were discovered, connecting the Malden Mills fire protection system to the fire protection system in an adjoining property, which was once a part of the same complex. If this valve had been opened, it could have delivered approximately 4000 gpm capacity from a second pumping station into the Malden Mills system.

The precise age and design of the fire protection systems is unknown, since the records and engineering drawings were destroyed in the fire. Changing technology and production methods along

---

[3] Sprinkler systems in the buildings to the south, which were not involved in the fire, were also supplied by this system.

with building renovations and additions over the years necessitated many changes in the fire protection systems. Additional protection was provided for high hazard process areas within the complex, including the hopper room where the explosion is believed to have originated. The property owners were reported to have made ongoing efforts to maintain and upgrade the systems and to keep up with state of the art in fire protection for their thriving business. The Methuen Fire Department backed and supported these efforts.

Guard Service

Three guard stations controlled the primary access points to the complex and roving security guards patrolled the property. The main guard station was located on the south side of Stafford Street, across from the Dye House. A second guard station was located in the parking lot near Broadway Street, in front of the Marriner Building, and the third was at the north end of the alley between the Process and the Main Buildings.

The main guard station also housed the on-site medical services and radio communications, as well as the fire alarm annunciator panels. When an alarm was received, maintenance workers and guards were dispatched by radio to investigate. The guard at the main station would call the fire department to confirm the alarm and provide a status report from the personnel on scene as additional information was available.

The fire department would respond to the main guard station and relied on the security guards to lead them to the problem area. The fire department did not have a written pre-fire plan for the complex, but did conduct familiarity inspections at the site from time to time.

Employees were instructed to move toward the main guard station during an emergency and to seek treatment for injuries at the first aid room. The company had an evacuation plan, which required supervisors to account for their employees in the event of an emergency.

Process Hazards

While the buildings in the Malden Mills complex were old, the processes and equipment inside were modern and involved advanced manufacturing technologies. These processes also introduced new materials and new challenges related to the operation and protection of the facility. Because the processes were sensitive to temperature, humidity and other environmental conditions, many of the window openings had been covered with plywood in an effort to better control the climate inside the buildings. The buildings were heavily loaded with materials that were used or produced, including tons of synthetic fabrics, drums of adhesives and several hazardous chemicals that were used to treat the materials.

There were three manufacturing "flock" lines in the middle section of the Monomac Building, two on the first floor and one on the second floor, making a popular insulating fabric known as Polartec. This product consists of a base fabric upon which treated nylon fibers are affixed, standing on end, to make a type of pile fabric. This material was produced in large quantities to meet the demand for lightweight winter clothing.

Two flock lines were in operation on the evening the incident occurred. The production lines were designed to operate as a continuous process. A stream of base fabric came off a roll and receiving a coating of a latex based adhesive, then moved into the 60 foot long hopper room where the nylon fibers were applied, then entered a 100 foot long dryer where the adhesive was set.

In the hopper room the short nylon fibers were applied to the fabric, floating down onto the adhesive as the fabric moved under a high voltage (up to 50 kV) electrical grid. The fibers were chemically treated to five them an inherent electrical affinity and, as they approached base fabric, the electric field would orient the fibers, causing them to land with only one end touching the adhesive. Different chemical treatments were used, depending on the specific type of material that was being produced. Although cleaning was an ongoing operation, the atmosphere in the area of the flock lines was dusty, because of the tiny fibers, and employees working in the area were provided with dust masks.

The dryers had to be precisely controlled to maintain the desired temperature of 450 degrees Fahrenheit for the proper time to set the adhesive. The fabric ran over a series of rollers weaving through a system of heated pipes, while a special heat transfer fluid was circulated through the pipes. Three large natural gas-fired heaters were used to heat the fluid, one connected to each dryer. The temperature of the fluid was monitored manually and only designated employees were authorized to adjust the settings on the heaters, which were located in a boiler room that had been added to the west side of the building.

The heat transfer fluid utilized in the process had been identified in an insurance industry report as a factor in 49 fires and 5 explosions over a previous ten year period. However, mill management was unaware of the potential threat posed by the use of the heat transfer fluid. The fluid, in the free burning state, produces a heavy black carbonaceous cloud of smoke that is prone to ignite, allowing the fire to spread rapidly.

The tiny nylon fibers were occasionally ignited as they passed through the electrical grid. Three different means were provided to shut down the power grid, stop the production line and initiate a stream flood to suppress a potential explosion within the machinery. A system of light weight nylon line was used as the first line fire detector – if a small fire or "sparkler" developed within the electric grid this line would melt very quickly and initiate the automatic response. An ultraviolet detection system was installed as a second means of activation and employees could also press a manual emergency button to achieve the same results. These systems were tested monthly, however, activation of the steam flood system did not transmit an alarm to the fire department.

Previous Incidents

The Methuen Fire Department responded to about forty fire alarms per year at the Malden Mills complex in recent years. These incidents ranged from dumpster fires and water flow alarms to a major incident in 1993, when an explosion in the hopper room of flock line number one seriously burned several workers. No fire ensued from the explosion and a multi-casualty incident (MCI) command structure was established to manage patient treatment and transportation. The cause of this explosion was determined to be a new treatment that made the fibers prone to ignition.

INCIDENT SUMMARY

Explosion

On the evening of December 11, 1995 at about 8:04 p.m., an explosion occurred in the center section of the Monomac Building on the first floor. Employees reported hearing and feeling a rumbling before flames erupted in the area of flock lines one and two. The guards at the main guard station heard the explosion, which was quickly followed by a water flow alarm. Responding security personnel found several injured employees, many severely burned, being helped out of the building by co-workers.

The Methuen Fire Department dispatched a normal first alarm response of 13 personnel to the waterflow alarm:

Car 2	Deputy Chief and driver
Engine 3	3 crew members
Engine 4	3 crew members
Rescue 1	2 crew members
Ladder 1	1 crew member
Ambulance 2	2 crew members

The waterflow alarm was quickly followed by a call from the main guard station at about 8:07 p.m., advising that an explosion had occurred and several workers were seriously injured. This information was relayed to the responding units while they were still en route. Methuen Fire Alarm quickly became overloaded with additional calls reporting the fire.

When advised of this additional information, the Deputy Chief (C-2) requested the dispatch of the two remaining Methuen companies, Engines 5 and 6, and mutual aid from Salem and Lawrence, including companies to cover the empty Methuen stations. Ambulance 2 notified CMED of the report of multiple burn patients and need for ALS. The first due paramedic unit, P-2 was dispatched by CMED and advised of the report from Ambulance 2. The paramedic in charge of this unit immediately requested the response of 10 additional ambulances and 2 medical helicopters.

Search and Rescue

Arriving on the scene at about 8:08 p.m., the Deputy Chief was told that workers were still trapped in the Monomac Building. Engine 4 and Rescue 1 were assigned to initiate search and rescue and, as they followed a guard to the alley between the Monomac and Process Buildings (Figure 5), they encountered several burned workers who were being assisted toward the guard station. Arriving at the scene, Ambulance 2 set up a multi-casualty incident (MCI) command post at the guard station and began triage.

The plywood coverings on the first and second floor windows of the Monomac Building had been blown off and the alley was littered with debris from the explosion. The interior was filled with smoke and the force of the wind was blowing the smoke out the first floor windows. Heavy smoke was also blowing over the top of the building, however, no fire was observed or reported at this point. A fire command post was established by C-2 at the southeast corner of the Monomac Building.

Employees directed the firefighters to a door about 100 feet from the north end of the building, where two crews donned SCBA and made entry to conduct search and rescue operations. Inside they found that the electricity had been knocked out by the explosion and water from activated sprinkler heads and broken pipes was cascading down on them. Engine 4 reported blowing smoke inside and some heat on the second floor, but did not see any fire inside the building.

Engine 3 arrived at about 8:12 p.m. and was assigned to conduct search and rescue operations in the center section of the building. Engine 5 was assigned to search the south end of the building. The driver of Engine 5, who was informed that an injured man needed help inside, left his apparatus, entered the building, and was able to remove 2 injured workers with the help of other employees.

Several employees were assisted out of the building and the interior search was completed before any of the firefighters had consumed a full tank of air. The Deputy Chief was informed that supervisors had accounted for all employees of the Mill.

Fire Suppression

At about the same time the search was completed, security personnel advised the Deputy Chief of a fire on the west side of the Monomac Building and the officer in charge of Engine 3 requested an engine to come to the northwest corner of the building. When the Deputy Chief went around the south end of the building to observe conditions on the west side, he observed fire around the loading dock and extending above the boiler room that housed the oil heaters. The flames were two to three stories high and 20 to 30 feet wide and a 40 mph wind was pushing the heat, flames and smoke into the building on at least three floors. The seat of the fire was behind several forty-foot long trailers and at least two of the trailers were already involved.

The magnitude of the situation was evident and the Deputy Chief quickly initiated the first of numerous requests for mutual aid to the Haverhill Communications Center. Access to the west side of the building was extremely limited, as most of the open area between the building and the river was occupied by more parked trailers. This left very little room to maneuver apparatus or to establish safe operating positions in the area where the fire was discovered. The strong winds, temperatures under 10 degrees Fahrenheit and a covering of snow added to the problems.

Engine 6 had stopped at the central fire station to pick up the Methuen Fire Department's second ambulance and was arriving at the scene at about 8:20 p.m., as the fire was reported to C-2. Moving around the south end of the building to the west side, Engine 6 was able to get close enough to initiate an attack on the fire with a 1-3/4-inch pre-connected line, but did not have a supply line from a hydrant. Engine 5 was directed by C-2 to locate a hydrant and to lay parallel 3 inch supply lines to Engine 6 for master stream operations. Engine 5 laid the lines across a small bridge to a hydrant on the opposite side of the river, however, the hydrant was frozen and inoperable.

At this point the Chief of the Methuen Fire Department (C-1) arrived, followed closely by Engine 7 and Ladder 5 of the Lawrence Fire Department. The 1-3/4-inch line was having no affect on the fire and Engine 6 was running out of tank water. Witnesses said the fire was moving across the west face of the building as fast as a man could walk. The Fire Chief ordered Lawrence Ladder 5 to take the place of Engine 6 and to set up a heavy elevated stream to the fire in an attempt to keep the fire from spreading. Lawrence Engine 7 located a yard hydrant, about 200 feet south of the building, but some of the trailers had to be moved to reach it. A 4-inch line was eventually connected to this hydrant to allow a master stream device to be placed in service.

The Fire Chief then went to the northwest corner of the Monomac Building where he met the crews that had completed the interior search. He ordered Engine 3 and Ladder 1 to set up a ladder pipe operation to protect the Acadia Mills Building, 50 feet to the north. Two 3-inch lines were stretched 200 feet from Engine 3 to a yard hydrant in the alley between the two buildings. Water pressure was good as this operation was established, but the fire was already pushing out of the Monomac Buildings into the alley and involving the enclosed passageway that spanned the alley between the two buildings. The ladder pipe had been in operation only a short time when the passageway collapsed across the supply lines. This occurred at about 8:45 p.m.

After all personnel in the area were accounted for, Ladder 1 was repositioned outside the collapse zone. The Monomac Building was now heavily involved in fire. Engine 3 was repositioned close to the river and the crew chopped through the thick ice, but their apparatus was unable to draft. Engine 4 was then moved into position to draft and was able to restore the water supply to Ladder 1. Engine 6, which had been moved to make room for Lawrence Ladder 5, breached the ice on the east side of the river and also attempted to draft.

Due to the rapidly expanding size of the incident and the limited resources, the Fire Chief took responsibility of the north side along Chase Street, while the Deputy Chief directed operations on the south side, along Stafford Street. They were attempting to find some way to surround the fire and contain it before it reached the boundaries of the complex. Both officers determined that additional assets were necessary immediately if they were to make any progress in stopping the fire and requested additional mutual aid.

Explosions from inside the Monomac Building broke through the west wall, showering debris in the area where Engine 6 was attempting to set up a drafting operation. The decision was made to withdraw Lawrence Ladder 5, which was not yet in operation, to the south end of the building. Engine 6 was relocated outside the collapse zone, where a second hole was made in the ice. Engine 6 made several attempts to draft from this opening, but ice build-up on the hard suction hose made this impossible. Engine 5 was then moved into position and drafted from this opening to supply a master stream and a hand line. Lawrence Engine 7 was also moved away from the building and joined Engine 5 across the bridge.

By 8:45 p.m. the wind was also pushing the flames from the Monomac Building across the alley on the east side and the fire was threatening to spread to the Process Building. Lawrence Ladder 5 was positioned to attempt to keep the fire from spreading across the alley, but did not have an adequate water supply to protect the exposure. Efforts were made to set up additional master streams as the wind pushed the flames directly toward the Process Building, however, there was not enough space between the buildings and not enough time or water to stop the progress of the fire.

The four sprinkler systems in the Monomac Building were overwhelmed and the water supply in the on-site system, which also provided the water for the yard hydrants, was severely compromised. The valves to shut down the sprinklers in the Monomac Building were in the alley that was severely exposed to the fire. The Mill's fire protection personnel tried to close section valves to isolate the compromised areas, but met with little success. The electricity for the complex was disconnected at about 8:50 p.m. disabling the electric fire pump, which reduce the water supply for the sprinklers and yard hydrants by about 50 percent.

By about 9:00 p.m., the fire had extended into the Process Building. As the Monomac Building began to collapse, the 40 mile per hour winds blew huge embers up to one and one half miles away. Mill security personnel had evacuated the buildings to the east in the path of the fire, while police officers were evacuating families and businesses between the mill complex and Broadway Street.

Several crews entered the Main Mill Building and attempted to use the stand-pipes to keep the fire from extending into this structure, but here was not enough pressure present to fill their hoses. With no fire department connections to augment the water supply, the interior crews had to evacuate as radiant heat from the advancing fire was causing window frames to ignite. The fire quickly spread into and through this building, then extended east into the Dye Building and the New Building.

Containment

By 9:30 p.m., a command post was set up in the area of the main guard station and several sectors were established to direct operations in different areas. A multi-agency command structure was set up with representatives from the State Fire Marshal's office, the senior officers of several fire departments and other agencies from the cities of Methuen and Lawrence. The strong wind was pushing the fire rapidly to the east, through the large mill buildings toward Broadway. The progress of the flames could not be stopped until the fire reached an open area that could be used as a fire break and enough resources could be assembled to set-up heavy defensive streams.

French Street was identified as the strategic location to stop the progress of the fire, as the east wall of the Main Mill Building had a setback of about 200 feet from this street. At the south end of this block, the space between the Main Mill Building and French Street was a parking lot, while the one story New Building occupied the north end of the block. Elevated streams were set-up to protect the homes on the east side of French Street and try to knock down some of the embers that were sailing overhead.

As mutual aid companies arrived, relays were established from both the Lawrence and Methuen water systems to supply master streams and drafting operations were set-up in at least four different locations along the river. Several mutual aid companies were also assigned to patrol the neighborhood downwind for fires started by the flying brands.

A second stand was established at the northwest corner of the Process Building where it connected to the Acadia Mills Building. Several companies were positioned to keep the fire from extending in this direction, supplied water from a drafting operation 600 feet to the north. Although it was severely exposed, the Acadia Mills Building suffered relatively minor damage.

A corrugated cardboard box manufacturing plant was located directly north of the Main Building, separated by only 50 feet. Radiant heat and embers ignited a fire on the roof that damaged this building and its contents, but the building was ultimately saved.

A light covering of snow on roofs and the ground extinguished many of the flying brands and embers as they landed in the area east of the main fire. However, at about 12:21 a.m. a fire was reported in a furniture refinishing company on Annis Street, about 3 miles downwind in a densely built area. The equivalent of a second alarm response was required to control this fire, which damaged several adjacent structures.

The fire was confined within 6 hours and under control approximately 12 hours after the explosion occurred, however, overhaul operations at the scene continued for several days.

Mutual Aid Response

The mutual aid response to this incident involved 434 personnel from 55 fire departments: 26 from Massachusetts and 29 from New Hampshire. A total of 55 pumpers, 15 ladder companies, 13 rescue units and numerous other vehicles and command officers responded to assist the Methuen Fire Department. Staging was established at Broadway and Center Streets, one-half mile north of the fire. A secondary staging area was established in Salem, New Hampshire for the large number of units responding from this direction.

EMS Operations

Methuen Fire Department Ambulance 2, with two EMTs on board, arrived at 8:07 p.m. and established a multi-casualty Incident (MCI) command post at the main guard station. As they began to triage the numerous injured workers who were arriving there, Ambulance 2 advised the responding ALS unit (P-2) they had 15 to 20 injured, some with serious burns.

At 8:13 p.m., P-2 arrived and one paramedic became EMS Command while the other took over Triage Officer responsibilities. They quickly requested more ambulances and helicopters, in addition to their initial request for 10 ambulances and 2 helicopters. In addition to dispatching additional EMS units, CMED contacted the area hospitals, as well as Regional burn centers and trauma centers. The hospitals were told of the situation and asked the number of critical care beds available, so that CMED would be able to assign transporting ambulances to hospitals with the proper resources available.

The EMS response included 25 ambulances, many from the fire departments in Massachusetts and New Hampshire that also assisted with fire suppression. Four medical helicopters were used to transport the most critically injured patients to burn centers in the Boston area.

As additional EMS resources arrived, they were assigned to treatment and staging positions. The ambulance staging area was designated as the south side of Stafford Street east of the guard station. This allowed for the passage of other emergency vehicles to conduct fire suppression operations as the MCI was in full swing. The main guard station was used for triage and treatment, as it was indoors and away from the cold weather and the fire scene. As patients were ready for transport, the next ambulance would drive up the east side of the building to load, then exit through the parking area to Broadway Street.

The first critically burned patients to be transported left the scene by ambulance at 8:32 p.m. At 8:41 p.m. the first helicopter arrived and a landing zone was set up in the parking lot immediately north of the guard station. Due to wind-driven smoke obscuring the landing zone, all subsequent helicopter transports were made from local hospitals to burn and trauma centers after the patients had been removed from the scene by ambulance. No other helicopters landed at the scene.

Most of the EMS activity at the scene took place during the first hour, as all of the injured employees were transported from the scene within 67 minutes of the initial call. A total of 27 employees were transported, 12 with critical burns. During this time one injured firefighter was also treated and transported. The EMS command structure and triage operation were winding down while the fire suppression operation was still expanding.

After all of the injured employees had been transported, the EMS command structure was reassigned to manage the Rehabilitation Sector. Rehab operated in the main guard station until 11:20 p.m., then moved to the southeast corner of the parking area to get farther away from the fire. A Red Cross canteen was also located in the parking lot. Eleven ambulances and one ALS unit were held to assist with Rehab and several relatively minor firefighter injuries were treated.

Four ambulances and their personnel were also used to assist with evacuation of the residential area immediately east of the fire area and an ambulance was sent to the Annis Street exposure fire to stand-by.

Overhaul

The overhaul phase continued for several days before all of the fire suppression resources could be released. Heavy equipment had to be brought in to move debris and remove hazards as several large buildings were reduced to rubble and the entire site was encased in ice and covered with snow.

Specialists had to be brought to the scene to conduct a survey for hazardous materials and potential sources of contamination. Several areas were identified within the rubble where hazardous materials were used or stored, including radioactive sources that were used in instrumentation in the process control systems. Runoff of products into the river was a major concern as melting occurred.

Investigation

Investigation of the explosion and subsequent fire began before the flames were under control. Due to the magnitude of the direct loss and economic impact on the community, the number of serious injuries and the sizes of the site, the investigation became a complex operation with many participants. The Massachusetts State Fire Marshal assigned a team to manage the overall investigation process, with individuals from several other agencies assisting and cooperating, including a strike team from the Bureau of Alcohol, Tobacco and Firearms. In addition, experts from the property and business insurers and independent investigators with expertise in specific areas were brought in to work with the investigation team.

The investigation continued for several months, partially due to the problem of excavating the site to search for evidence. The rubble of several large buildings, including boilers and heavy machinery, was covered with ice and snow. As the weather permitted, the site was methodically cleared, searching for clues that would help to identify the cause of the initial explosion and explain the rapid fire spread throughout the complex.

While there were many theories and dozens of witness statements, there was no clear explanation of where and how the explosion originated. The flocking process that occurred in the hopper room was implicated, however, there was also evidence to suggest that the origin was related to the dryers or the oil heating system. Different witnesses provided different observations; none could fully explain the sequence of events that occurred. Most of the witnesses agreed that there was no indication of a problem before the initial explosion occurred.

In addition to searching of the cause of the explosion, the construction of the buildings and the details of the fire protection systems were analyzed and documented. Most of the plans were destroyed in the fire and many of the details of the system arrangements could only be obtained from the careful examination of the ruins over several months of investigation.

At the time this report was prepared, the official cause of the explosion had not been determined.

Lessons Learned And Reinforced

1. Pre-fire planning is an essential fire department function

Pre-fire plan information is particularly important for large complex properties. Gathering and reviewing pre-fire information helps firefighters to become familiar with the locations where they are likely to face challenging situations. The information should be used to evaluate fire department readiness and resource capabilities, identify potential problems and hazards and formulate strategies for predictable situations. At the fire scene, the pre-fire plan information can be used to provide essential information for the Incident Commander to support strategic planning.

Pre-fire plans for large complexes should identify building and fire protection features which are likely to be significant in a fire or emergency situation and record the information in a format that can be used at the scene of an incident. Although the wind driven flame front spreading from the Process Building to the Main Building prevented access to control valves, detailed information about the arrangement of the sprinkler and water supply systems could have been extremely valuable at this incident. The Incident Commander would have had a better understanding of the fire protection systems and control points.

2. Limited resources result in limited operational capabilities.

The Methuen Fire Department was severely handicapped by the limited resources that were available during the early stages of this incident. This included the number of companies responding on the initial alarm and the number of personnel assigned to each company, While more than 400 firefighters eventually responded, the resources available during the first few minutes, when multiple priorities had to be addressed, were very limited. Search and rescue clearly took priority over fire suppression efforts during those first critical minutes.

3. Protection of large fire risk properties depends on automatic sprinklers

Automatic sprinklers were invented to protect this type of complex and still offer the only reliable and effective means to deal with the tremendous fire risk of mill construction industrial properties. In this incident, the automatic sprinkler system was compromised by the initial explosion and rendered ineffective. Once the fire fully involved the Monomac Building due to the explosion of the flock fibers, the ignition of the heat transfer fluid and the 40 mile per hour winds, it was almost impossible to stop the spread of fire through the remainder of the complex.

4. Fire departments must take an active part in fire prevention and code enforcement.

The property owners had taken an active approach to fire protection, with full-time personnel responsible for maintaining and upgrading fire protection equipment and systems. However, fire department connections were not provided to supplement the water supply for sprinklers and standpipes, as there was no code requirement in the State of Massachusetts requiring mill-type buildings to retrofit their systems with fire department connections. In addition, storage in trailers blocked access to fire hydrants and limited apparatus access to the fire area. Fire departments should take an active part in verifying that all of the installed fire protection systems meet code requirements. In this case, the insurance companies performed several fire safety inspections each year and the fire department relied heavily on these inspection reports.

5. An effective incident management system is essential for large incidents

This incident presented tremendous challenges to the fire departments and other agencies that responded. The need for an effective system to direct and coordinate an operation of this magnitude and complexity is evident. This was particularly difficult during the early stages, when the first arriving command officers had to deal with multiple priorities and rapidly changing conditions. The complexity of the management challenge increased as the huge mutual aid response was assembled, further emphasizing the value of formal ICS practices.

6. Proper risk assessment made this fire relatively safe for firefighters

This fire presented the incident commander and senior officers with a monumental risk management problem. After search and rescue had been completed there were no lives to be saved, but there was a huge property value at risk, involving the community's largest employer. There was an added risk that the high winds could spread the fire beyond the complex to other parts of the city. A series of decisions had to be made to determine where and when to attempt to stop the fire and the degree of risk that was appropriate for firefighters, considering the limited resources that were available during the early stages.

The fact that only six relatively minor injuries were reported by firefighters suggests that the risk management decisions were appropriate.

7. Major incidents require expanded communications capabilities

The sizes and complexity of this incident, including the huge mutual aid response, overwhelmed the communications systems that were available. The Methuen Fire Department's radio and telephone capabilities were burdened quickly and the need to obtain additional resources placed heavy demands on the mutual aid coordination center.

The handheld radios used by Methuen command officers could not communicate with other agencies. The lack of common radio channels was a major problem until command vehicles with all of the different frequencies could be assembled at the command post. The situation delayed utilization of some of the available resources and hampered effective coordination.

APPENDIX A

DIAGRAMS

Figure 1. Area Plan

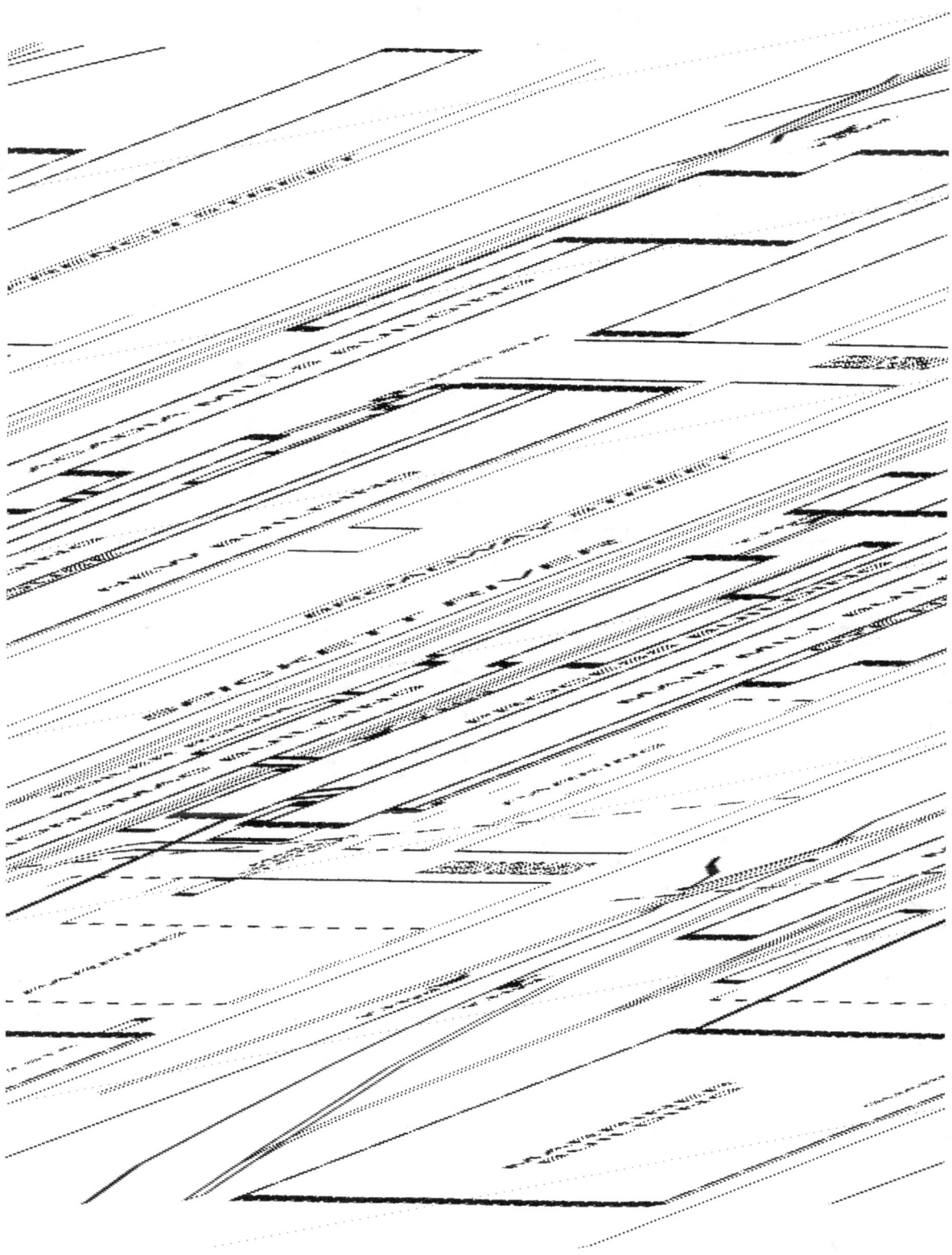

Figure 2. Site Plan

Figure 3. Private Water System

Figure 4. Municipal Water System

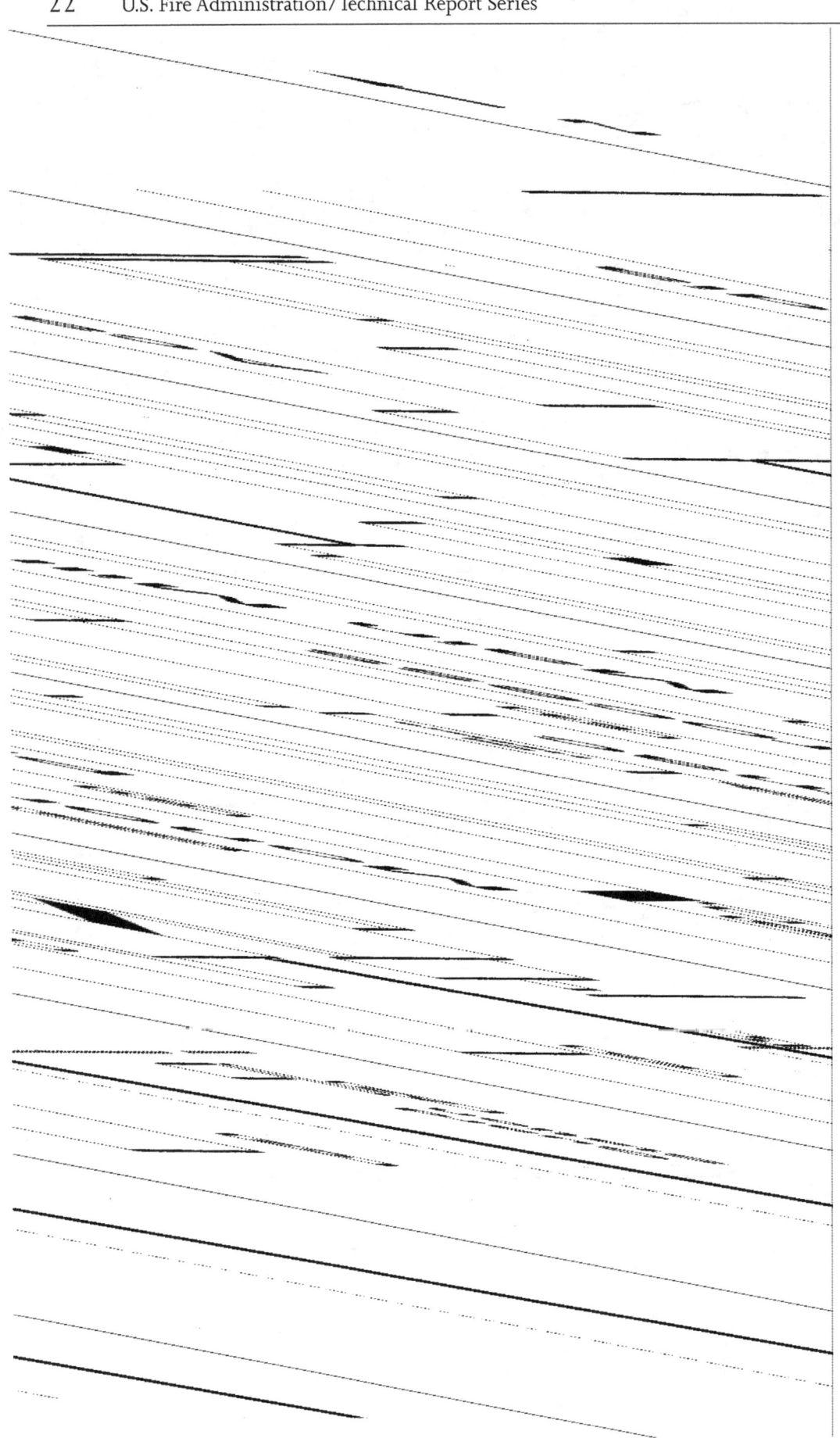

Figure 5. Placement of Initial Response for Search and Rescue

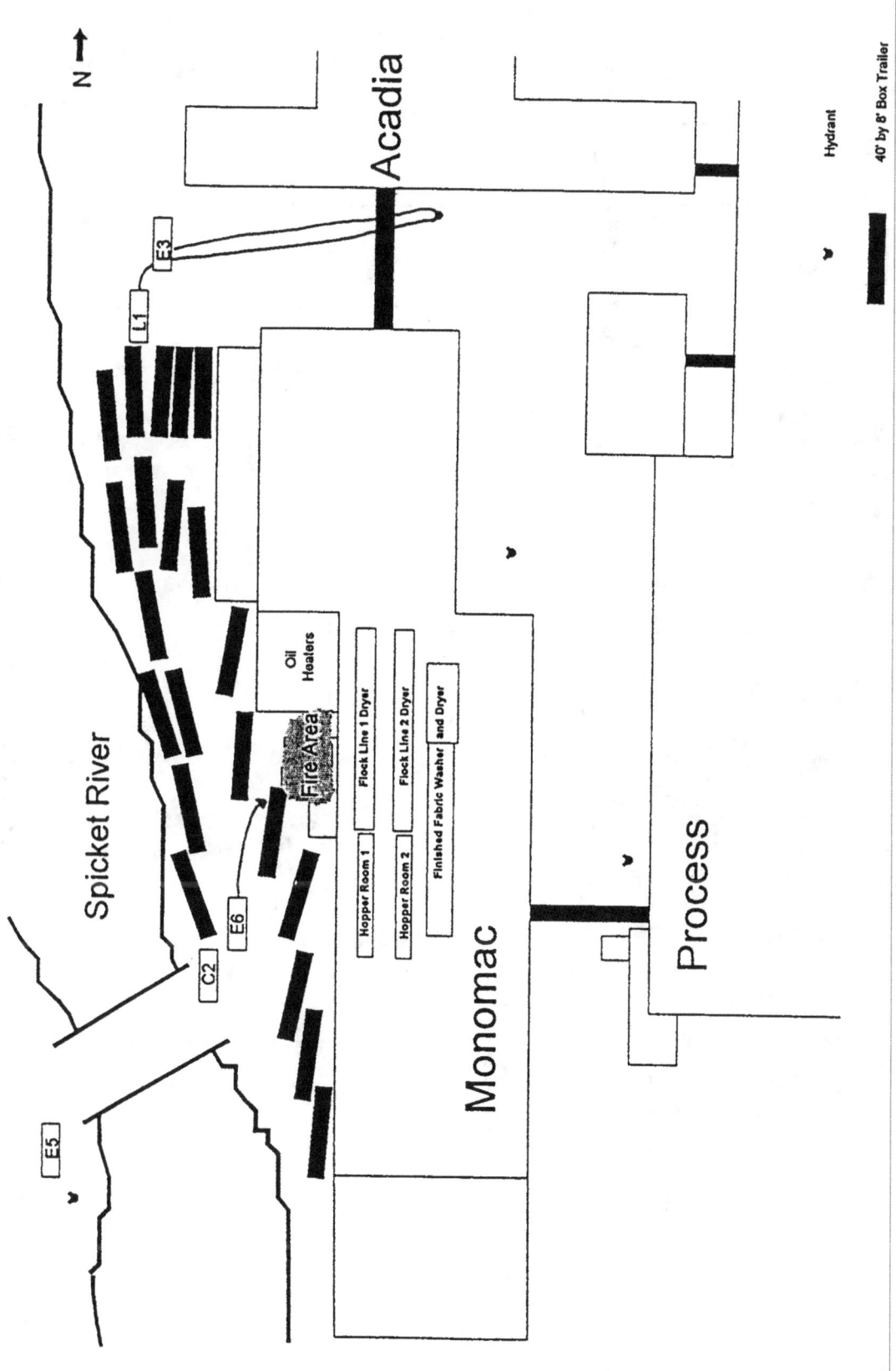

Figure 6. Apparatus Placement at approximately 8:30

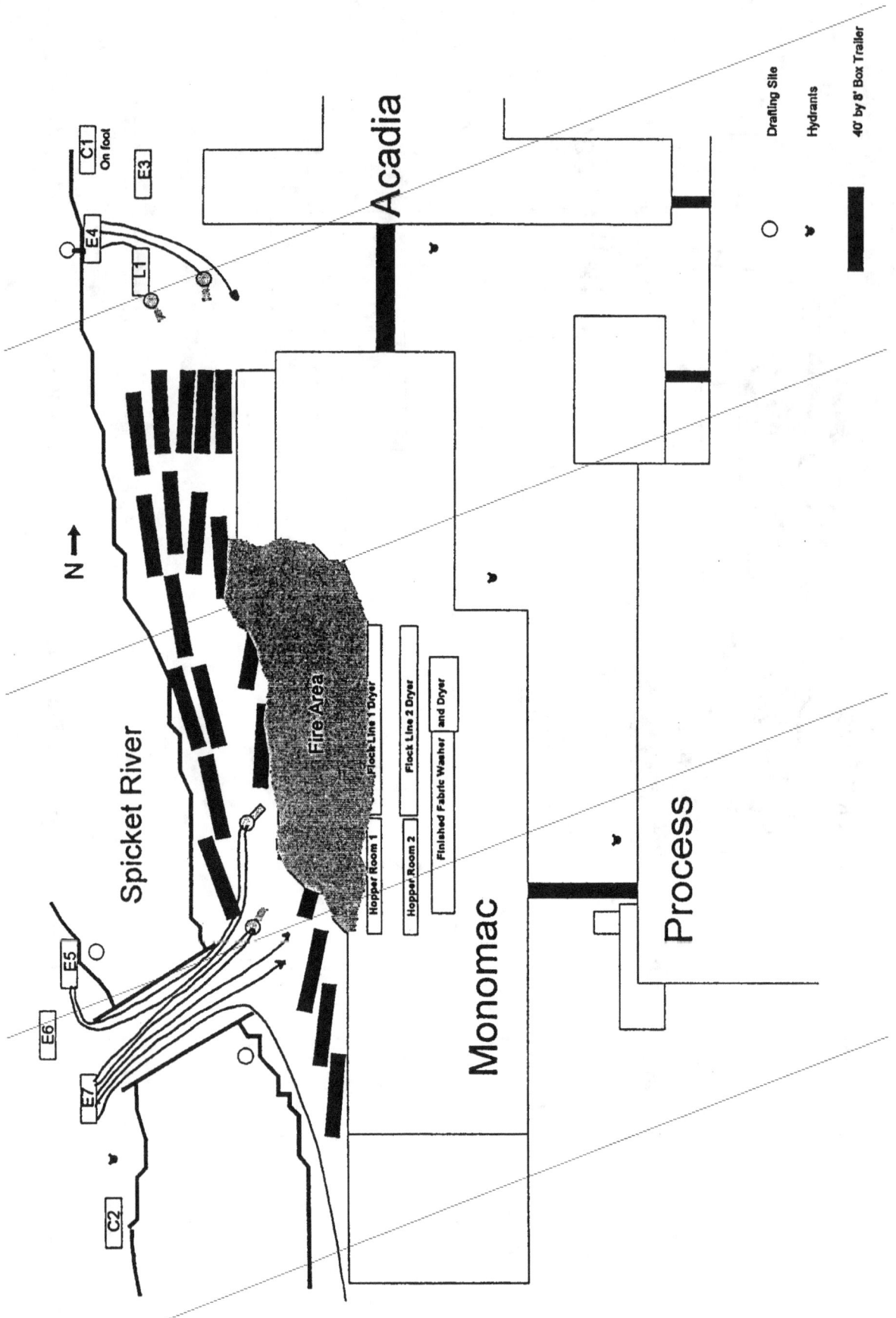

Figure 7. Apparatus Placement at approximately 8:50

APPENDIX B

PHOTOGRAPHS

Photographs 1 to 3 were captured from video taken during the fire. The video tape was supplied by the Methuen Fire Department and the Massachusetts State Police. The balance of the photographs were taken by Scott Howell. Color versions of the video captured photographs and others in this report can be obtained from the USFA's WEB page.

1. Ladder pipe operations near the northwest corner of the Monomac Building early in the fire. Vantage point was across the Spickett River.

2. The fully involved west side of the Monomac Building; floors have collapsed as have some walls. Vantage point was across the Spickett River.

3. Aerial view of the fire scene early afternoon on December 12, 1995 looking southwest. The Mariner and Van Brodie Buildings are in the background.

4. Mop-up operations on the west side of the site. The bridge over the Spickett River is in the background. Note the tight conditions and truck trailer parking.

5. Remains of the Monomac Building looking to the south.

6. Conditions on the west side of the site with the truck trailers and the Spickett River.

7. Composite of the overhead connecting bridge between the Monomac and Process buildings.

8. Example of the undamaged overhead connecting bridge in another part of the site.

9. Typical mill building construction with heavy wood beams and thick wood floors.

10. Example of the heavy buildings that were involved in the fire.

11. Windows once needed for natural light have been covered over, making visible fire location difficult. In addition, depending on the materials used, ventilation may be difficult.

12. An idea of the scale of the combustible mill buildings that were involved in this incident.